GENETICS

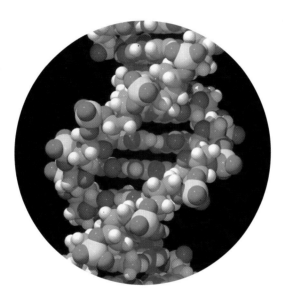

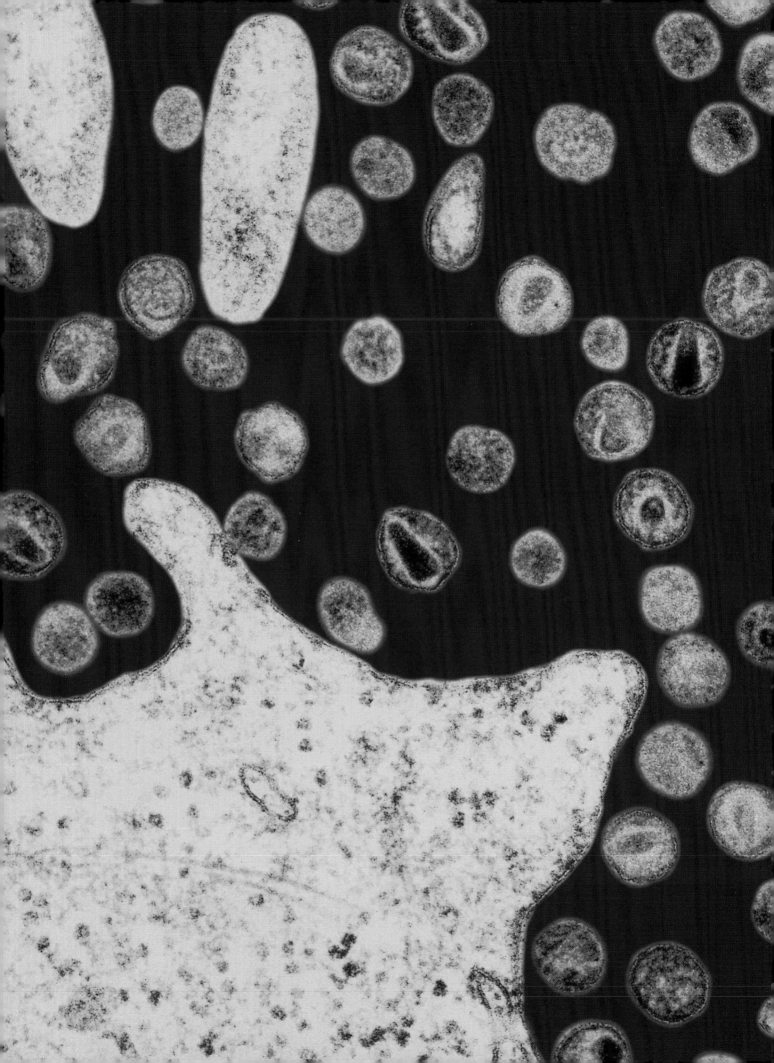

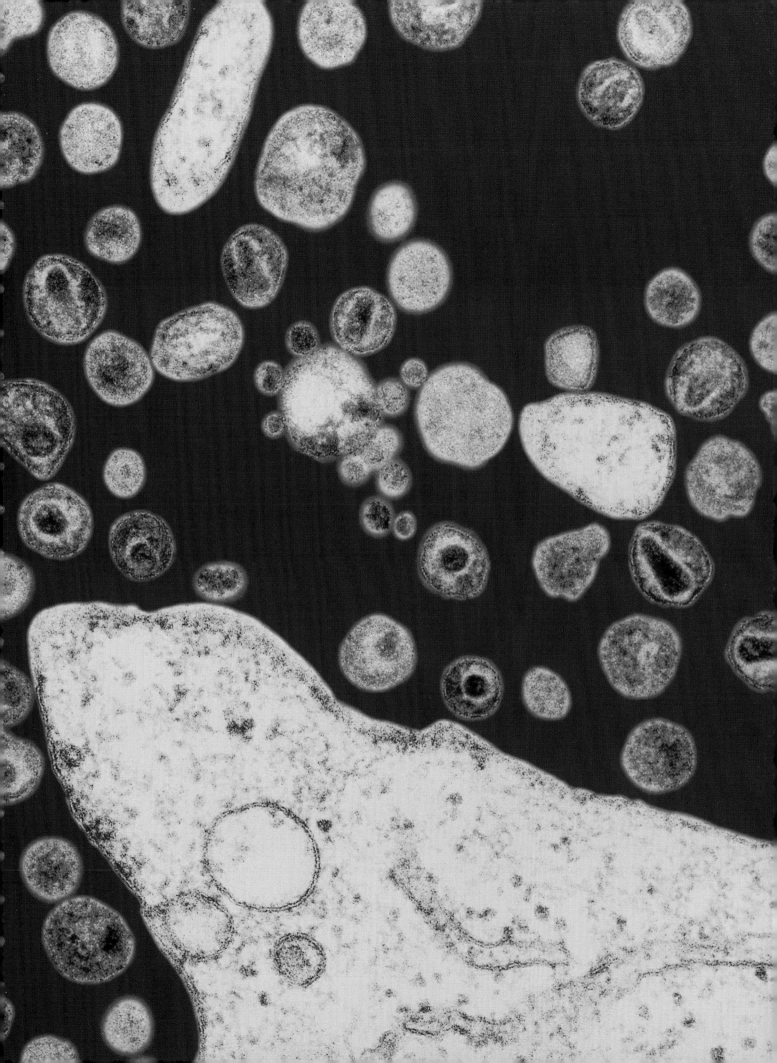

GENETICS

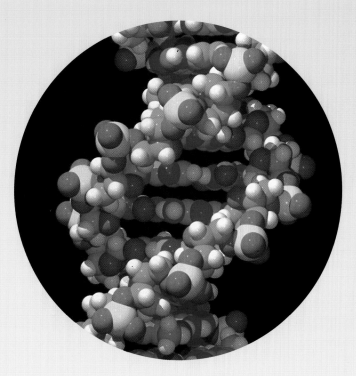

Richard Beatty

RAINTREE
STECK-VAUGHN
PUBLISHERS

A Harcourt Company

Austin • New York
www.steck-vaughn.com

Produced by Roger Coote Publishing

Published by Raintree Steck-Vaughn, an imprint of Steck-Vaughn Company

Design and typesetting	Katrina ffiske and Victoria Webb
Commissioning Editor	Lisa Edwards
Editors	Alex Edmonds and Jon Ingoldby
Editorial Consultant	Steve Parker
Picture Researcher	Lynda Lines
Illustrator	Michael Posen
Consultant	Dr. David Cove

Raintree Steck-Vaughn Staff:

Editor	Sean Dolan
Project Manager	Max Brinkmann

Library of Congress Cataloging-in-Publication Data

Beatty, Richard.
   Genetics / Richard Beatty and Steve Parker
      p. cm. — (Science fact files)
   Includes bibliographical references and index.
   ISBN 0-7398-1015-4
      1. Genetics—Juvenile literature. [1. Genetics.] I. Title. II. Series.

QH437.5 .B43 2001
576.5—dc21   00-045695

Endpaper picture: Electron microscope picture of HIV virus
Title page picture: Computer graphic of DNA molecule

We are grateful to the following for permission to reproduce photographs: Bruce Coleman Collection 8 bottom (Pacific Stock), 29 top (Staffan Widstrand); NHPA 9 bottom (GI Bernard); MPM Images 23 top, 28 top, 39 top; Oxford Scientific Films 10 bottom (Deni Bown), 14 (Kent Wood), 23 bottom (Alison Kuiter), 31 top (Michael Leach), 31 bottom (W Lummer/Okapia), 34 (Science Pictures Ltd); Science and Society Picture Library/Science Museum 32, 40 top; Science Photo Library front cover left (Will & Deni McIntyre), front cover top (Rosenfeld Images Ltd), 5 (Ken Edwards), 8 top (James King-Holmes), 10 top, 12 top, 13 (Dept of Clinical Cytogenetics, Addenbrookes Hospital), 15 top (Quest), 15 bottom (Quest), 16 (Bryson Biomedical Illustrations, Custom Medical Stock Photo), 17 top (Andrew Syred), 18 (Ken Edwards), 19 top (Maximilian Stock Ltd), 19 bottom (A Barrington Brown), 20 top (Philippe Plailly/Eurelios), 20 bottom (Eye of Science), 22 (Profs P Motta and T Naguro), 24 (Jerry Mason), 25 (Dr Linda Stannard, UCT), 27 top (Eye of Science), 27 bottom (Darwin Dale), 28 bottom (Novosti), 30 (George Bernard), 33 top (Sinclair Stammers), 33 bottom (Philippe Plailly), 35 top (Matt Meadows/Peter Arnold Inc), 35 bottom (NIBSC), 37 (James Holmes/Cellmark Diagnostics), 38 (Catherine Pouedras/Eurelios), 40 bottom (Christian Jegou/Publiphoto Diffusion), 41 top (Rosenfeld Images Ltd), 41 middle (Philippe Plailly/Eurelios), 41 bottom (Philippe Plailly/Eurelios), 42 left (Simon Fraser), 43 top (Matt Meadows/Peter Arnold Inc), 43 bottom (Will & Deni McIntyre); Stock Market 9 top (Ariel Skelley), 26 (Lester Lefkowitz), 29 bottom, 36, 40 middle (Fernando Bertuzzi); Stone front cover right (Mark Joseph), front endpaper (Hans Gelderblom), 11 top (Michael Heissner), 12 bottom (Peter J Bryant/BPS), 42 right (Spike Walker); Topham Picturepoint 39 bottom (Peter Jordan).

The statistics given in this book are the most up to date available at the time of going to press.

Printed in Hong Kong by Wing King Tong

0 1 2 3 4 5 6 7 8 9   WKT   05 04 03 02 01 00

CONTENTS

The words that are explained in the glossary are printed
in **bold** the first time they are mentioned in the text.

INTRODUCTION

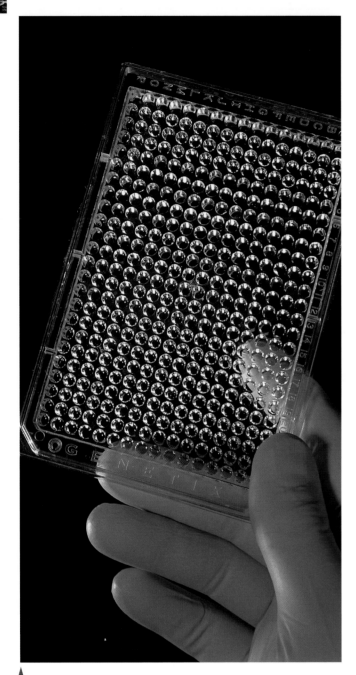

When living things breed, or reproduce, they make more of their own kind. A mother lion gives birth to a baby lion—not to a baby tiger or a baby cheetah. A maple seed grows into a maple tree, not into an oak or elm tree.

In addition, the baby lion does not look like just any other lion. It closely resembles its mother and father. The same happens with all offspring. They resemble their parents more than they resemble other members of their kind, or **species**. This tendency of offspring to resemble their parents is called **heredity**. Heredity is one of the main subjects of the branch of science known as genetics.

Instructions for Life

Genetics is the study of **genes**, what they are, how they work, and how they can be changed. Genes are the instructions for how a living thing grows, develops and carries out life processes inside its body. These instructions are inside living things in the form of a chemical called **DNA**, deoxyribonucleic acid.

Geneticists are scientists who study inheritance and genes—especially the chemical that makes up the genes, which is called DNA. This tray contains fragments of human DNA.

The tendency of children to look like their parents is part of heredity.

Our varied appearances are partly due to the genes passed from our parents.

Genetics is one of the newest, fastest-moving and most exciting areas of science. It is hugely important in many ways—from farming, food production and wildlife conservation to health, medicine and catching criminals.

Controversies

Genetics is involved in various topical arguments such as genetically modified crops, cloning animals and even creating completely new forms of life. This is partly because genetics is developing so rapidly. Its methods and techniques progress quickly and are becoming extremely powerful. It is difficult for ordinary people to keep pace with research, to understand what genes are and how they are changed, and to be aware of what might happen in the future.

FUTURE FILE

GENETIC RESEARCH: GOOD OR BAD?

Some people support genetic research. They say that it can lead to many positive results, such as ending hunger around the world and fighting illness. With precautions and safeguards, the risk of problems is tiny. Other people say that genetic research is too hazardous and moving too fast. It should be halted until we know more about the risks. They fear that it may bring terrible dangers, such as new and untreatable diseases. However, one matter seems certain—genetics is likely to become even more important in the future.

Could genetics bring back the quagga, a type of zebra extinct since the 1880s?

HOW GENETICS BEGAN

nlike many sciences, the study of genetics can be traced back to a specific time and place. This was the middle of the 19th century in the town of Brno, now in the Czech Republic.

Gregor Mendel (1822-84) was a monk who also taught science and was interested in gardening. In his spare time at his monastery, he grew garden peas to study how they changed from one generation to the next—parent to offspring. The pea plants differed in small ways. Some grew tall, others stayed short. Some had white flowers, others were purple. Some produced smooth seeds, others grew wrinkled seeds. Mendel recorded all these features, called characters or traits, for each pea plant as he worked.

Years of Peas

Mendel carried out a series of tests, breeding together certain pea plants. With a small brush he carried the pollen grains from the male part of one flower to the female part of another flower. In this way he knew exactly which pea plants were the parents. He did this hundreds of times, took the seeds that formed, grew them into pea plants, studied their features, and bred them too. He recorded all the details in his notebooks, growing a total of more than 21,000 pea plants.

Gregor Mendel began his pea-growing experiments in about 1853.

Mendel studied various physical features of peas, such as flower color.

FACT FILE

WHEN FEATURES ARE LINKED

Mendel's work showed that some inherited features or characters of pea plants were not affected by others. They were passed on independently. For example, there was no tendency for wrinkled seeds to be inherited along with purple flowers. But other features were likely to occur together and be passed on with each other. This is known as "linkage" (see page 17).

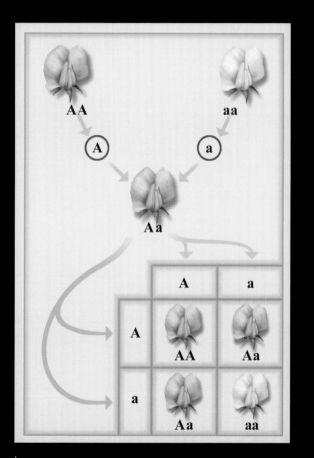

A white-flowered and a purple-flowered pea plant do not produce equal numbers of white- and purple-flowered offspring, due to dominance (see page 23).

Male pattern hair loss (baldness) is a feature that tends to run in families.

Not Blended Together

At the time most people had a vague idea about inheritance. In general, offspring looked like their parents. A child looked like its mother and father. But it was widely believed that the parents' features combined or blended in the offspring. So the offspring should look like an in-between or halfway version of the two parents.

Mendel's work overturned this traditional view. For example, he found that breeding a pea plant with smooth seeds and one with wrinkled seeds did not result in a pea plant with part-wrinkled seeds. The seeds were either fully smooth or fully wrinkled. This seemed to happen for other features too.

Hereditary Factors

Mendel's experiments showed that many features, instead of blending, passed unchanged from parent to offspring. He suggested that some unknown items inside living things controlled their inherited features. He called them hereditary factors or inherited particles. We now call them genes.

Mendel's work laid the basis for the science of genetics. But it caused little interest at the time. It was only taken up by other scientists in the early 1900s, almost 20 years after his death.

CELLS AND CHROMOSOMES

From about 1900 scientists looked again at the work of Gregor Mendel. They carried out more breeding experiments and used microscopes to peer into living things, to find genes. All living things are made up of tiny **cells** that can only be seen through a microscope. Each cell is like a building block or "unit of life," able to grow, carry out life processes, take in nourishment, survive, and split into two to make more cells.

Some living things are single cells, like the bacteria and other "germs" that cause disease. Other living things, like maple trees, lions, and humans, are made of billions of cells.

Sets of Genes

Experiments showed that genes are inside every cell. What is more, a living thing made of many cells has the full set of its genes in each of its cells. However, not all the genes are used in every cell. There are many different types of cells in the body, such as skin cells, blood cells, and muscle cells. Each type of cell is specialized to do a certain job and has its own shape and way of working. It uses only some of the genes it contains. These are "switched on" while the rest of its full set of genes are "switched off."

Thomas Hunt Morgan (1866–1945) discovered genes are contained in chromosomes.

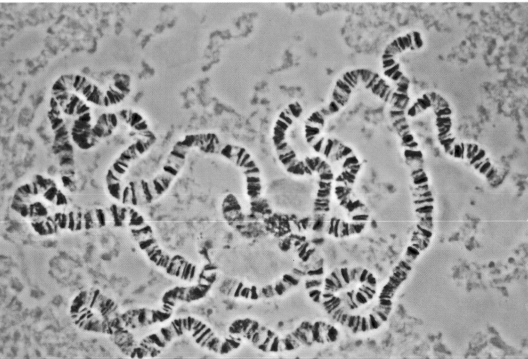

The microscope reveals that certain types of chromosomes, like these from the fruit fly, have stripes or bands.

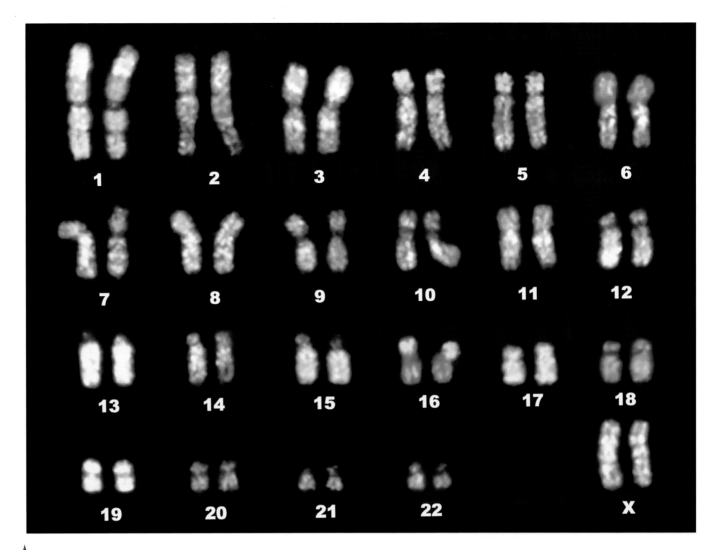

The full set of chromosomes from a living thing is its karyotype. This is the human karyotype of 46 chromosomes.

Spotlight on Chromosomes

Further experiments showed that the full set of genes in each cell is in one part, the control center, or nucleus, of the cell. They are contained in thread-like parts inside the nucleus called **chromosomes**.

A typical human body cell has 46 of these chromosomes. They exist as 23 pairs. One chromosome of each pair came originally from the mother and the other in the pair from the father. Cells from other living things have different numbers of pairs. Some flies, for instance, have eight chromosomes—four pairs in each cell.

Each chromosome is like a long, microscopic string or necklace of thousands of genes. Chromosomes are seen most clearly when a cell is about to split or divide into two, when they look like microscopic X or Y shapes.

 **FACT FILE**

BOY OR GIRL?
In each set of 46 human chromosomes, 2 are called sex chromosomes. This is because they carry instructions for whether the body is female or male. In a girl both chromosomes are the same, known as X chromosomes (XX, as shown above). In a boy there is one X chromosome and another smaller one, the Y chromosome (XY). Most animals and plants have sex chromosomes. They work in a similar way to human sex chromosomes, although they differ in size and shape. Birds, for example, are the opposite of humans. The male has two X chromosomes (XX) while the female has an X and a Y (XY).

HOW GENES ARE PASSED ON

In a living thing such as a plant, animal, or human, the full set of genes is present in every body cell, on thread-like structures called chromosomes (see previous page). A basic feature of all life is that cells divide to make more cells. A single cell divides to form two offspring cells. This is how single-celled life-forms such as bacteria reproduce. It is also how larger, more complicated life-forms such as ourselves grow and develop. And it is how we repair and maintain our bodies as cells constantly divide to replace old, worn-out, or damaged parts.

How a cell divides.

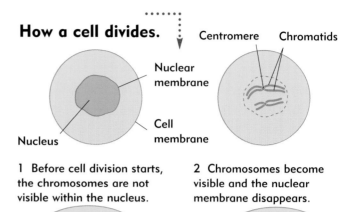
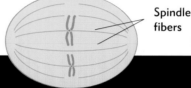

Nuclear membrane

Centromere   Chromatids

Nucleus   Cell membrane

1 Before cell division starts, the chromosomes are not visible within the nucleus.

2 Chromosomes become visible and the nuclear membrane disappears.

Spindle fibers

3 A kind of "scaffolding," called the mitotic spindle, forms. The chromatids shorten and become attached to the spindle.

4 The paired chromatids split and are pulled to opposite ends of the cell. Each chromatid is now a separate chromosome.

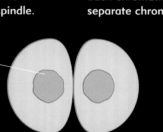

Chromosome is too thin to see

5 New nuclear membranes form, the chromosomes uncoil and become less visible, and the cell divides into two offspring cells.

FACT FILE

HOW CELLS DIVIDE

A typical cell splits or divides in a series of stages, called phases. Unless some error occurs, the two offspring cells are identical to each other and to the parent cell.

1 **Interphase** Chromosomes are in the cell's nucleus but are too long and thin to see clearly. Each chromosome copies or duplicates itself (by DNA replication, see page 19). The original chromosome and its new identical partner are **chromatids**.

2 **Prophase** The chromatids coil up to become shorter, thicker and more visible. Each pair is joined at a point along their length, the centromere.

3 **Metaphase** The membrane or "skin" of the nucleus breaks down and the pairs of chromatids line up in the middle of the cell.

4 **Anaphase** The chromatid pairs split at their centromeres. The two members of each pair separate and move to opposite ends of the cell.

5 **Telophase** The chromatids have now become chromosomes. Each group gathers at one end of the cell and the cell pinches itself into two offspring cells.

This cell from an onion plant's root is in the anaphase stage of mitosis.

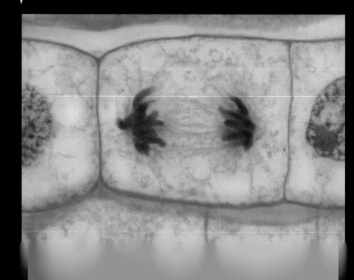

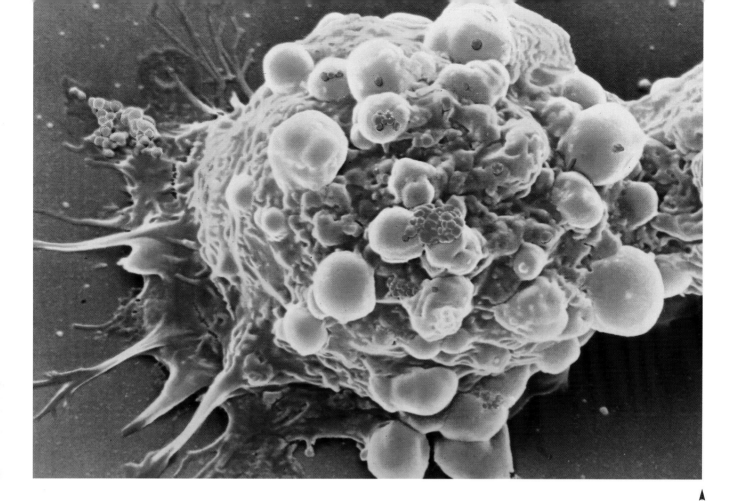

Copying Genes

When a cell divides, somehow the chromosomes with their genes must be copied or duplicated. This ensures both offspring cells receive the complete set of genes. This type of division happens in the cell's nucleus with its chromosomes and is called mitosis. It is followed by splitting of the whole cell. The whole process occurs in a series of stages as shown opposite. The stages ensure that the chromosomes with their genes are copied correctly and shared out equally between the two offspring cells.

Control of Cell Division

Cell division is controlled by various chemical substances within the cell. Some encourage or promote it, others prevent or inhibit it. These substances, like all other chemicals in the cell, are made under the instructions of genes. If one of these genes becomes altered it may not work properly and produce its substance to control cell division. Cells could then divide in an uncontrolled way and spread through the body. This is what happens in certain kinds of diseases called cancers. For this reason, a huge research effort continues in order to understand cell division better.

Cancer cells like this breast cancer cell (with blue blobs on its surface) often divide out of control.

A cancer cell in the last stage of division by mitosis. The two offspring cells are almost separate, joined only by a narrow "bridge" of spindle fibers.

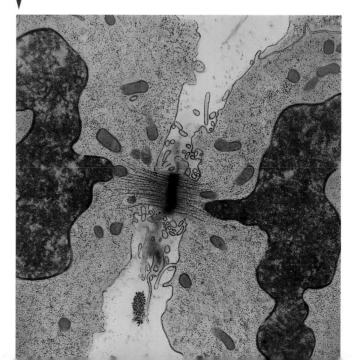

HALVES AND DOUBLES

New cells for body growth, development, maintenance, and repair are produced by splitting, or mitosis. Reproduction occurs when two sex cells, an egg cell from the mother and a sperm cell from the father, come together. They join to form a single cell, the fertilized egg. This begins to divide into many cells (by mitosis) and develop into a new individual.

Halving the Number of Chromosomes

If egg and sperm each had the full number of chromosomes—for example, 46 in a human—the fertilized egg would have double this number, 92. Doubling of the chromosome number during reproduction is prevented by a special type of division, meiosis. The cells made by meiosis develop into sex cells—eggs in a female or sperm in a male.

The 46 human chromosomes are in 23 pairs. In

The type of cell division called mitosis, shown on the previous page, results in two offspring cells. Meiosis, shown above, produces four offspring cells (along the bottom of the picture) from one original parent cell (top right). Each of these offspring cells has only one of each chromosome, not a pair.

meiosis only one member of each pair passes on to the egg or sperm. So these sex cells each have 23 chromosomes. When an egg and sperm join, the normal number of 46 chromosomes is restored in the fertilized egg for the newly developing individual.

Geneticists say that a normal body cell, with

Each tiny pollen grain from the male parts of a flower, in this case a daisy, contains one male sex cell made by meiosis. Pollen is carried by animals or blown by wind to the female parts of a flower of the same kind.

chromosomes in pairs, is **diploid**. An egg or sperm, with only one of each chromosome, is **haploid**. The human diploid number is 46, that is, 23 pairs of chromosomes. The human haploid number is 23.

Inherited Together

The fact that genes are grouped together on chromosomes explains one puzzle of inheritance. This is linkage—when features or traits tend to be inherited together, rather than separately. This is because the genes for these features are on the same chromosome. So they are usually passed together from parent to offspring.

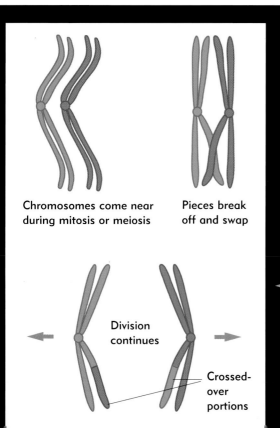

Chromosomes come near during mitosis or meiosis

Pieces break off and swap

Division continues

Crossed-over portions

FACT FILE

SINGLES AND PAIRS
Mosses and ferns look similar to other plants. But in their genetics they have a fundamental difference. Most other plants, like animals, are diploid. That is, each body cell has pairs of chromosomes. Mosses and ferns are the reverse. The main plant is haploid, that is, each of its cells contains just one of each chromosome.

When chromosomes come together at cell division sometimes they touch, break, swap pieces, and rejoin. Known as crossing-over, this brings together genes in new combinations as part of genetic diversity (see page 31).

WHAT ARE GENES MADE OF?

By the 1940s scientists were working to discover the chemical substances that make up genes. Chemical studies showed that chromosomes contain large amounts of substances called proteins. Could genes be made of proteins?

Proteins Galore

Proteins are among the most important substances in living things. There are thousands of kinds in two main groups, structural proteins and functional proteins.

Structural proteins are "building blocks." They make up body parts like skin, hair, muscles, and bones. They also form the tiny parts of cells, such as the cell membrane (the "skin," which encloses a cell).

Functional proteins are involved in the way the body works, and especially in controlling processes inside cells. Every second in a cell thousands of substances join and come apart in hundreds of chemical reactions. These reactions are controlled by functional proteins called enzymes.

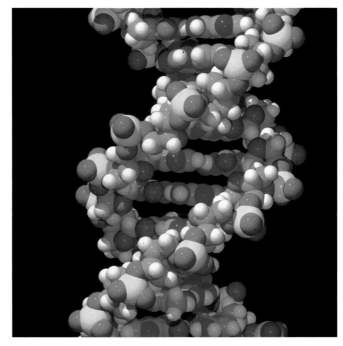

A computer model of DNA showing its twisted, or spiral, structure.

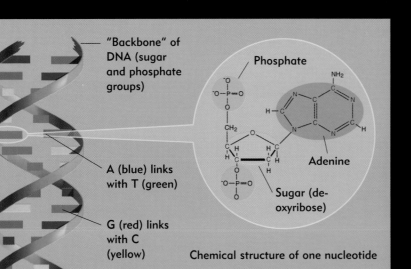

"Backbone" of DNA (sugar and phosphate groups)

Phosphate

A (blue) links with T (green)

G (red) links with C (yellow)

Adenine

Sugar (de-oxyribose)

Chemical structure of one nucleotide

The double helix of DNA has subunits called bases arranged in linked pairs. The base adenine, A, always links with thymine, T. Guanine, G, joins to cytosine, C.

FACT FILE

COPYING DNA

DNA has four kinds of bases—adenine, guanine, thymine, and cytosine, known as A, G, T and C. A always joins by its cross-link to T, and G to C. When a cell divides, the two DNA strands "unzip." Then a new strand is assembled as a partner for each existing one. This is called DNA replication. The new strand is an accurate copy because each base can only have one possible partner. An A on the existing strand must have a T on the new strand, and so on. This produces two identical double helixes of DNA

In a test tube, DNA resembles pale, jelly-like fibers.

📖 HISTORY FILE

DISCOVERING DNA'S SHAPE
The double helix shape of DNA was worked out in 1953 by scientists Francis Crick and James Watson at Cambridge University in England. They used scientific clues and evidence gathered by other scientists, including Maurice Wilkins and Rosalind Franklin at King's College in London. This discovery is now seen as one of the most important in the history of science. Watson, Crick, and Wilkins shared a Nobel Prize in 1962.

James Watson (left) and Francis Crick explain their model of the double helix structure of DNA.

The Molecule of Life

However, experiments have shown that genes are not made of proteins. They are made of a different substance found in chromosomes. This is called deoxyribonucleic acid, DNA.

DNA is long and thin. In fact, in each chromosome there is one piece or molecule of DNA that is several centimeters in length. This is immensely long for a molecule. But DNA is so thin that it can only be seen using a very powerful microscope. Also it is coiled and folded up so tightly, thousands of times, that the chromosome that contains it is also very small.

Structure of DNA

Each molecule of DNA has two long, ribbon-like strands that wind round each other in corkscrew fashion. This forms a structure known as a double helix. The strands are joined by cross-links so the whole structure looks like a twisted rope-ladder.

Each ribbon-like strand of DNA is made up of individual chemical units called nucleotides (shown opposite), strung together like beads on a necklace. Every nucleotide contains a subunit known as a base. This joins to, or pairs with, another base on the other DNA strand, by the cross-link—the "rung" of the DNA ladder. The identity of the bases makes up the genetic code, as shown on the next page.

HOW GENES WORK

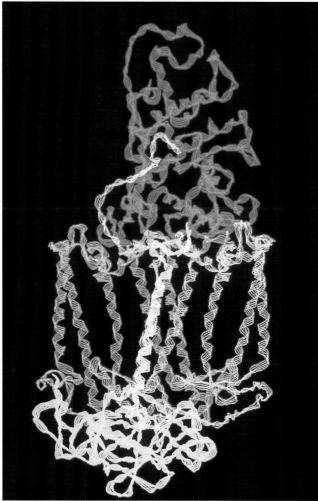

A gene on its own cannot do anything. It needs other parts of the cell to have its effect. In general, the effect of one gene is to make one protein. Thousands of different proteins are vital parts of living things (see page 18). Some form structural parts such as skin, hair, nails, bones, and muscles. Proteins called enzymes control the chemical processes and reactions inside the cell.

DNA to RNA

Making a protein from the information in the gene begins in a similar way to copying DNA, as shown on the previous page. First the two strands of the DNA double helix "unzip" along part of their length. This exposes some of DNA's bases A, T, G, and C. The order of the bases represents the information in the gene in chemical form. The bases are like letters in a code word—the genetic code (see Fact File on page 21).

One of the strands is used as a template to produce, not a partner strand of DNA, but a strand of the similar substance **RNA**, ribonucleic acid. RNA also has four bases. Three are the same as DNA's but instead of DNA's thymine, T, RNA has uracil, U. The copying of DNA into RNA takes place in the cell's nucleus and is known as transcription.

Many proteins are made of long chains folded into complicated shapes. This is a protein inside a bacterium, which traps the energy in light rays.

Chromosomes (here from a human cell) contain long strands of DNA and also structural proteins known as histones.

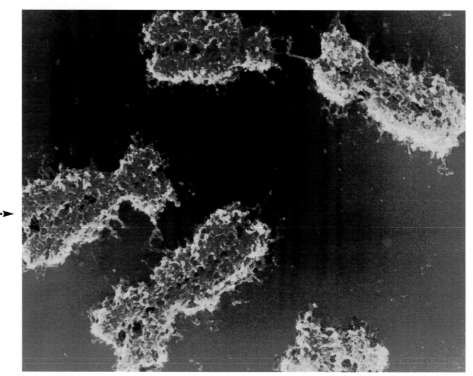

Ribosome moves
along RNA strand

Messenger RNA
was made using DNA

Order of bases in messenger RNA
determines order of amino acids

Ribosome

Transfer RNA
has given up
its amino acid

Amino acids joined
to protein chain

Transfer RNAs bring
amino acid subunits

**Building a protein involves
more than one type of
RNA. It takes place at a
tiny part of the cell called
a ribosome.**

Protein chain
lengthens

Protein Building Blocks

After transcription comes translation—making a
protein from the single strand of RNA. Like DNA and
RNA, each protein is a long molecule consisting of
many linked subunits. Protein subunits are known as
amino acids, and there are about 20 different kinds.
The order or sequence of amino acids is different for
each of the thousands of kinds of proteins.

RNA to Protein

The piece of RNA moves from the nucleus into the
main part of the cell. The order or sequence of its
bases determines which amino acids are clipped
together to make a particular kind of protein. In this
way information passes from the gene as DNA, to the
"go-between" of RNA, to the protein, which becomes
part of the body. For example, the gene for brown
eye color carries the information to make a brownish
protein, which gives the eyes their shade.

**Amino acids have different shapes and
join at different angles. Their order in a
protein gives the protein chain its shape.**

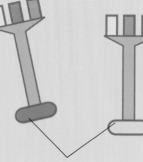
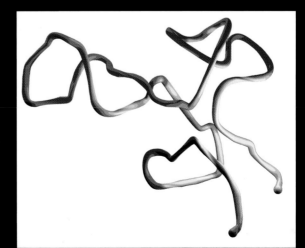

THE EFFECTS OF GENES

Genes affect many features of a living thing, such as its size, shape, the colors of its different parts, even how fast it grows and its resistance to disease. However, it is not quite as simple as "one gene controls one feature." Often one gene affects more than one feature. Also, one feature may be affected by several different genes.

In addition many features of a living thing are by things other than genes. They are also influenced by the environment—where it lives, what it eats, the weather, and other conditions. Often it is very difficult to decide on the relative effects of genes compared to the effects of the environment.

DNA is found in chromosomes inside a cell, and also in mitochondria. These are tiny sausage-shaped parts in a cell that break down food to provide energy. Mitochondrial DNA is much shorter than chromosomal DNA.

Different Forms of a Gene

In most animals and plants, chromosomes occur in pairs. The two members of the pair have equivalent genes. In a plant, for example, there may be a gene for flower color. In fact the plant has two of these genes for flower color. One is on the chromosome from one parent, the other on the chromosome from the other parent.

These genes may not be exactly the same on the two chromosomes. In many cases a gene exists as two or more different forms, known as **alleles**. For instance, the gene for flower color may have a red form or allele, which produces red flowers, and a yellow allele, which results in yellow flowers.

Strong and Weak Genes

So flower color depends on which forms, or alleles, of the gene were inherited. If both chromosomes have the red form or allele of the gene, the plant develops red flowers. Similarly if both have yellow alleles, the result is yellow flowers. (When both alleles are the same like this, a living thing is said to be homozygous for that particular gene.)

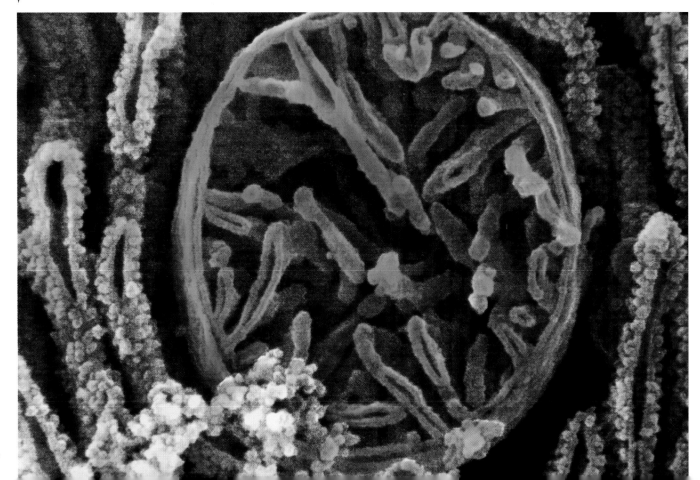

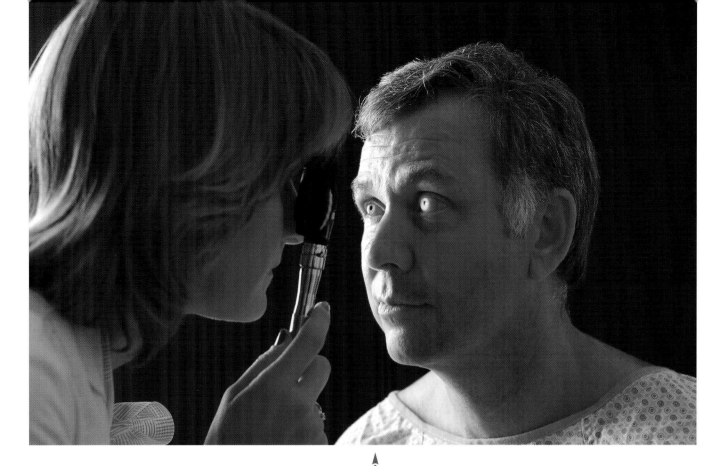

Dominant and Recessive

What if the plant has a red allele from one parent and a yellow allele from the other parent? Often the result is not a combination or blending of the two—in this example, reddish-yellow flowers. Instead one allele is "stronger," or **dominant**. It has its full effect and overpowers the "weaker" or **recessive** allele. If the red allele is dominant over the yellow one, the plant

In people the main gene for eye color has various versions, or alleles. Some alleles are dominant over others.

grows only red flowers. (When two alleles are different like this, a living thing is said to be heterozygous for that particular gene.)

FACT FILE

NAMING GENES

Gene names are usually printed in italic (sloping) type, *like this*. Many genes are named after the effects they have if they are faulty. For example, in most animals early in life, one main gene carries the information to "switch on" development of the eyes. If the gene is faulty, the animal does not develop eyes. So the gene is called *eyeless*. Some genes only "switch on" at a certain temperature (right).

Genes for dark fur only work in cooler parts of a white Himalayan guinea pig.

CONTROLLING GENES

Each living thing carries a full set of all its genes in all of the cells in its body. But all of the genes cannot be active in all of the cells. Cells are different and carry out different tasks because only some of their genes are "switched on."

Living things have several ways of controlling the activity of their genes. In some cases a cell makes extra copies of a particular gene, in the form of extra lengths of DNA. These multiple copies give the gene a greater or more powerful effect.

Switching Genes

A more common type of control happens when a gene, in the form of DNA, is copied to make first a strand of RNA and then a protein. Near to the main part of the gene is a region of DNA called a promoter. Here a type of control protein, called an enzyme, attaches in order to make the RNA strand. In many cases, however, the enzyme cannot attach on its own. Another type of control protein, called a transcription factor, must also be present. Each gene may have its own type of enzyme and transcription factor. If one or both are missing, the RNA strand cannot be made and so the gene is inactive—"switched off."

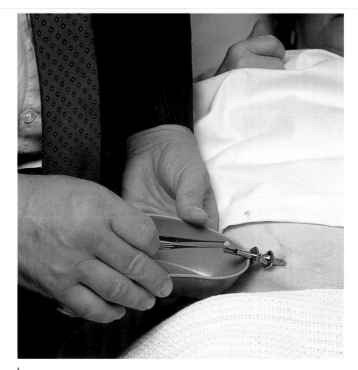

When the female reproductive cycle ceases at the menopause, hormone replacement therapy (HRT) can help to keep genes active and reduce problems.

Various additional substances or "extras" may be needed before a gene can become active and be used to make RNA for building proteins.

Length of DNA

Transcription factor

Transcription factor attachment site

Enzyme attachment site (promoter region)

Enzyme

Direction enzyme will move when transcribing gene into RNA

DNA in main gene zone can be used as a template and transcribed into RNA

MAIN GENE ZONE OF DNA

Chains of Command

The enzymes and transcription factors that allow genes to work are proteins. They are produced by instructions in other genes. In this way it is possible for one gene to affect another: genes control other genes, which control yet more genes. The genes are arranged in groups, which have different levels of importance. Switching on a single "command gene" may trigger a long and complicated series of events, in the way that an order from a basketball coach is acted on by all the members of the team.

Switches From the Outside

Certain substances in the body can also affect genes. These include hormones, which are "chemical messengers" carried around in the blood. They control many body processes, such as growth and response to stress. A cell may already contain control proteins that switch on various genes. But the control proteins only become active when a particular hormone enters the cell and joins onto them. The hormone and control protein together switch on their genes.

HISTORY FILE

THE FIRST GENE SWITCH

Microscopic living things called bacteria, especially the type known as *E coli*, are often used in genetic studies. In 1960 French scientists François Jacob and Jacques Monod discovered how these bacteria switch on a gene. That allows them to use a particular kind of sugar, lactose (milk sugar). When the bacteria detected this sugar in their surroundings, a gene became active for breaking it down to obtain energy. Jacob, Monod, and colleague Andre Lwoff shared a Nobel Prize in 1965.

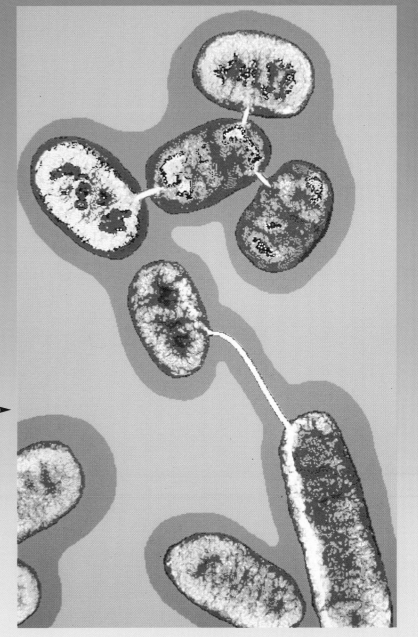

Escherichia coli bacteria are often used in genetic research. They are tiny, rod-like microbes found in many environments including the soil and inside the human digestive system. They can multiply by splitting in half, like other cells. Sometimes they exchange pieces of DNA along tubes (see page 34).

GENES AND DEVELOPMENT

One of life's greatest mysteries was how a tiny egg cell, no bigger than a pinhead, grows and develops into a huge plant or animal. Today it is known that the egg is "pre-programmed." It contains all the instructions or genes, in the form of DNA, to develop into that particular plant or animal.

Genes in the Egg

A new life begins as an egg cell from the mother is joined or fertilized by a sperm cell from the father. Both egg and sperm each contain a single full set of genes. The fertilized egg cell divides into two cells, then four cells, eight, and so on. The number of cells increases faster and faster to form a tiny developing embryo.

Genes in the Embryo

The embryo's genes direct its development. Some switch on others in a complex series of commands. Certain important control genes produce substances that affect the activity of other genes. The control substances seep or diffuse through the embryo, from one cell to the next. If a control substance is present in large amounts, at high concentration, it switches on a certain set of genes. Many cells away, where the control substance is weaker, different genes become active.

In a baby developing in the womb, different genes activate in different cells.

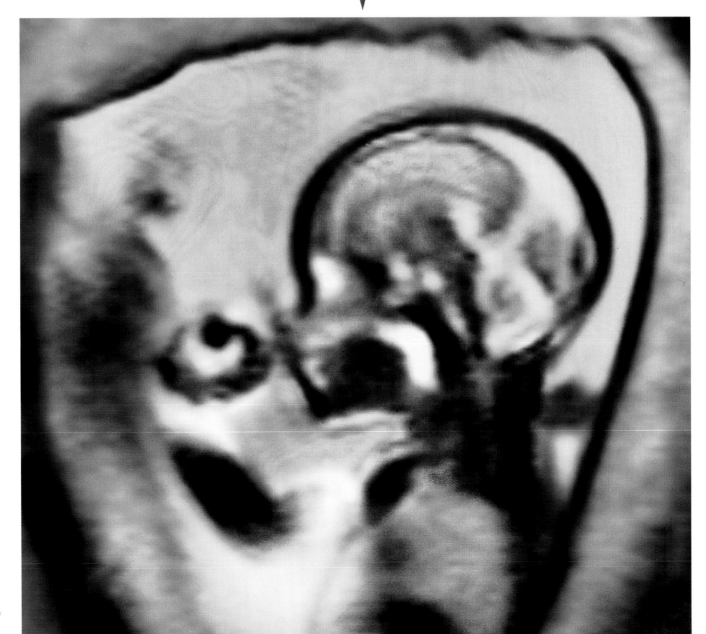

FACT FILE

GENETIC "GUINEA PIG"

Fruit flies, *Drosophila*, are frequently used for genetic experiments. They are small and easy to keep in the laboratory. They thrive on overripe bananas. They mate readily and breed quickly, so many generations can be studied in a year. They also have very large chromosomes that can be studied in detail under the microscope. Many major discoveries in genetics were first made using fruit flies.

................▶

In addition to its two large red eyes, this fruit fly has two small extra eyes instead of antennae (feelers).

There are many control substances, and each has its own effect on the development process. Gradually various genes switch on and off to alter patches of cells. The cells develop into specialized groups of similar cells, tissues, to form parts of the body such as the brain, eye, muscles, bones, and nerves.

Control Genes

The way that a fruit fly maggot (larva) changes and develops into an adult fly depends on eight major control genes, lying one behind the other on one of the fly's chromosomes. Different groups of cells in the maggot turn into the head, legs, wings, and other main parts of the adult.

Similar control genes have been identified in many other living things, from mice to humans. Plants develop in a slightly different way, but still under the influence of their genes.

The fruit fly has eight chromosomes, arranged in four pairs, in each of its body cells.

WHEN GENES CHANGE

Genes exist as lengths or sections of the substance called DNA. Hundreds or thousands of genes in a row form one immensely long piece or molecule of DNA. Molecules of DNA are "packaged" inside each cell as thread-like parts called chromosomes.

A cell from the human body contains 46 chromosomes (23 pairs). Each has a single length of DNA. Joined end to end, these 46 molecules of DNA would stretch about two meters (six feet). They only fit into the cell's nucleus because they are folded and coiled, and the folds and coils are also folded and coiled, so the great lengths of DNA pack into a very tiny space. Between them the 46 molecules of DNA carry the 100,000 or so genes needed for the human body.

All cheetahs have very similar genes. This means that cheetahs, as a species, are said to have a very small gene pool.

Radioactivity from the Chernobyl nuclear accident in 1986 caused mutations in living cells.

A hat and sunglasses help to shade the head from harmful UV rays.

FUTURE FILE

WARNING—TOO MUCH SUNSHINE!
In recent years global pollution has damaged the ozone layer. This layer of gases around the Earth prevents the surface from receiving too much ultraviolet light, UV, from the Sun. UV light causes sunburn and can also trigger mutations in skin cells. The cells may multiply out of control and form cancerous growths called melanomas. This is why too much direct sunshine and sunbathing can be harmful.

Mutations

When DNA copies itself, as cells divide, the process is usually very accurate. But sometimes an error occurs. A change in DNA is called a mutation. If it happens in ordinary body cells it will affect all the cells that result, but not any offspring of that individual person, animal, or plant. But if it happens during the making of egg or sperm cells, then the mutation and its effects will pass to the offspring.

Large and Small Mutations

Mutations happen fairly regularly in nature. They vary from an alteration in just one "letter" of the DNA code, to the appearance of extra whole chromosomes, even sets of chromosomes.

The results of mutations also vary greatly. Some have no effect. Some cause problems or damage. A few are helpful and have benefits. For example, a mutation in the genes controlling an animal's fur growth might make the hairs become longer. This could help the animal to survive in cold weather. Mutations are one source of the genetic variety important in evolution, as shown on the next page.

Making Mutations

Mutations can also be created in the laboratory. In 1927 genetics expert Hermann Muller showed that mutations could be caused in fruit flies by exposing them to X-rays. This was useful because the mutant flies could be used for genetic study. It also showed that X-rays damaged living things.

Other types of rays, such as ultraviolet light (see above), can also cause mutations. So can certain chemicals, such as those in cigarette smoke. These mutagenic chemicals may affect body cells and make them divide in an uncontrolled way, producing diseases such as cancer.

Different types or varieties of bananas are called "sports." They arise by change or mutation in the basic banana plant's genes.

GENETICS AND EVOLUTION

Evolution is the change in living things over time. Since living things grow and develop partly according to their genes, and many of their bodily features are controlled by genes, evolution can also be described in terms of genetics. Evolution is a change in the genes or genetic features in a group, or population, of living things over the course of generations.

Small and Large

Evolution often happens in a slow, gradual way. The wing color of a particular kind of moth may alter over many generations. This happens because the gene for the new wing color gradually spreads as the moths breed. Small differences like these can build up. After a time, usually hundreds of generations, the new moths look quite different from the original moths. A new species has evolved.

How Evolution Happens

For evolution to happen, a group of living things must have genetic variety. They all have the same basic genes, but different individuals possess different forms or alleles of the genes. In the moth example, the gene for wing color might exist in several forms. One produces light green wings that are well camouflaged on the light green leaves all around. These moths survive better than moths with other colors of wings. Occasionally a change or mutation in the gene produces different colored wings. But these moths show up clearly on light green leaves and are soon eaten by predators. Their unhelpful mutant genes disappear with them.

Worker ants all have very similar genes and help each other to survive.

HISTORY FILE

RAW MATERIAL FOR EVOLUTION

As genetics developed in the early 19th century, the genetic basis of evolution became more clear. A variety of genes is needed among living things for nature to select the best and "fittest" ones for conditions at the time. Mutations are one source of genetic variety. Also sometimes during cell division, chromosomes break along their length and join up with broken-off parts of different chromosomes. This process, called crossing-over (see page 17), brings together genes in new combinations and is another source of genetic variety.

Changing Conditions

But gradually, surroundings change. Plants with dark green leaves may become more common. Light-winged moths show up on them and attract predators. However, a genetic mutation crops up that produces dark green wings. Moths with this form of the gene are more likely to survive and pass it to their offspring. This is known as "survival of the fittest." Nature seems to choose which moths are most likely to survive—a process called natural selection. Gradually the allele for dark green wings spreads among the moths.

Coping with Change

Conditions such as climate are always changing on Earth. So living things must continually evolve to survive. Greater genetic variety in a group or population means that there may be some genes, somewhere, that help the group to survive as conditions change. Some farm crops and animals are so intensively bred that they have very little genetic variety. They cannot cope with a change such as a sudden drought or a new type of disease. These genetically limited breeds include various types of wheat, apples, cows, pigs, and chickens.

When the size of a group of living things is very small, such as voles on an island, a limited variety of genes can cause inbreeding problems.

Banded snails have a range of band widths and background shell colors. Each act as camouflage in a certain habitat.

31

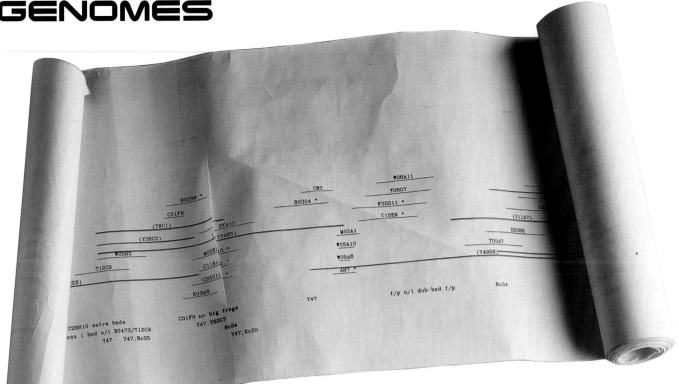

A single complete set of all the genes for a living thing is called its genome. The number of genes in a genome ranges from only three or four in the smallest, simplest forms of life—viruses—up to an estimated 100,000 or more for large animals, including humans.

Reading the Human Gene Map

The Human Genome Project was a massive international research project launched in 1990. Its aim was to identify all of the 3 billion DNA "letters" in the human genome. In June 2000 the project achieved its first main aim—the complete map of all the codes for an entire human being.

Identifying Genes

Many different methods were used to achieve this goal. The approximate positions of genes can be worked out by studying how they are inherited. For example, the more often that two genes are inherited together or linked, the closer they are on the same chromosome. Studying many such genes has produced maps showing the relative positions of genes on chromosomes. These act as a guide to build up more detailed maps of the DNA itself.

A map of chromosome number 2 of the tiny worm shown opposite, *Caenorhabditis elegans*. The map shows the relative positions of various genes and sequences of DNA. It is part of the research work that was done to identify all the genetic code letters in all of the worm's genes.

FACT FILE

JUNK DNA

Scientists were surprised to discover that only 3 percent (less than one-thirtieth) of the DNA in a human cell has actual codes for genes. The rest is made up of non-working copies of genes and other useless pieces of DNA. This spare, apparently useless 97 percent is sometimes known as "junk" DNA.

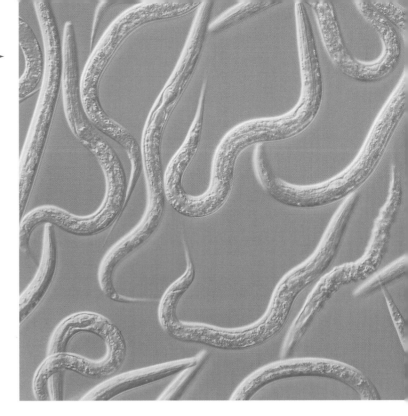

This tiny worm, *Caenorhabditis elegans*, was the first animal to have all its genes mapped.

Studying DNA

To examine the DNA, it was first broken into segments or fragments using chemicals. These segments were put into bacteria to see which proteins the bacteria made with them, so revealing the instructions in the gene. More chemicals and other techniques were then used to "sequence" each fragment, to work out the exact order of the DNA code letters.

The Tasks Ahead

More tasks on the human genome will occupy scientists well into the 21st century. The sequences of DNA code letters are now known. But which ones make up which genes? As each gene is located, will it be possible to work out the exact protein that it produces? What might that protein do in the body? And how do the 100,000 or so genes affect or control one another?

Genome research occupies thousands of researchers around the world.

Private Genes

There is special interest in genes that may cause disease. An individual person can have her or his genome studied to detect such genes. The genetic information for many people could be collected as a "DNA database." But who should have access to this information? A person known to be at risk of genetic disease might find greater difficulty getting employment or medical or life insurance. There are many such controversies about DNA databases and genetic information.

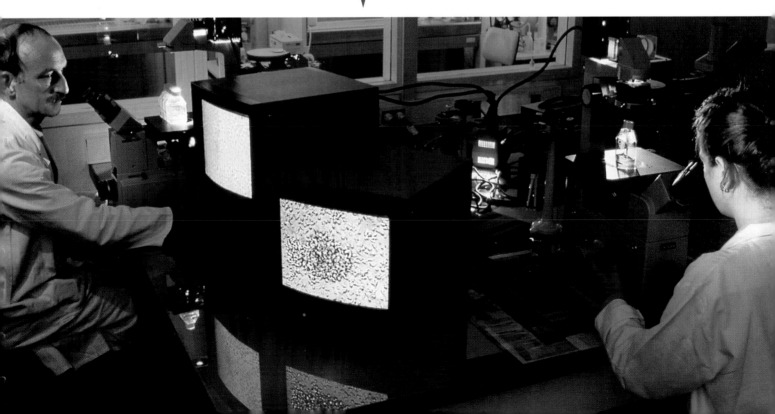

MICROBES AND GENES

Bacteria and viruses are microscopic, single-celled forms of life. They are found almost everywhere—in air, water, soil, and inside living things. Some bacteria and all viruses are "germs" that cause disease. But they have also been an enormous help in genetic studies. In particular they have been vital in developing techniques known as genetic engineering.

How Bacteria Live

Bacteria are smaller than the normal cells that make up the bodies of animals and plants. Also a bacterium has no nucleus (control center) or chromosomes inside. Instead there is a single hoop-like piece of DNA. This circular DNA molecule carries the genetic information. Bacteria, being so small and simple, have fewer genes than large animals or plants. The number ranges from less than 500 to around 4,000 depending on the type of bacterium.

When Bacteria Breed

Bacteria multiply fast, every 20-30 minutes, and can be grown in the billions in laboratory flasks. Usually they reproduce by splitting into two, as in ordinary cell division. But they can also transfer genes between themselves using small "extra" pieces of DNA called plasmids. These plasmids are used by genetic engineers to transfer genes from one kind of living thing to another, a process called genetic modification. It may help to treat genetic diseases and produce better-growing farm animals and crops.

The most commonly studied bacterium is *E. coli* (see page 25). It lives in many places, including inside the intestines of human beings, where it generally causes no harm. The entire genetic makeup of *E. coli* is now known. It contains about 4.5 million DNA code letters (compared to around 3 billion in a human being).

◀ This computer picture shows a bacteriophage or "phage," a type of virus that invades bacteria. It has an angular head, a stem, and long legs, almost like an insect— but it is millions of times smaller. In the background are the rounded bacteria that it attacks.

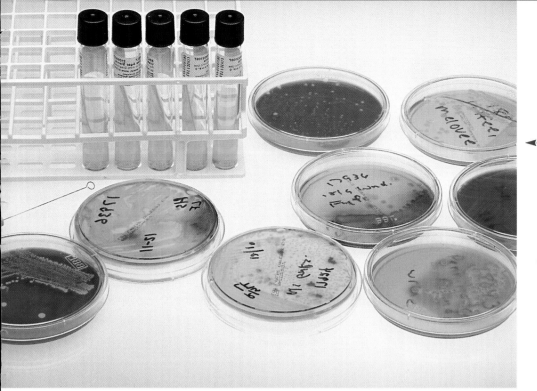

Bacteria can be grown quickly in huge numbers in laboratory containers, making them useful for genetic research.

Viruses

Viruses are even smaller and simpler than bacteria. Some have only three or four genes. Each virus consists of a tiny piece of DNA or RNA surrounded by a protective coat of protein. A virus multiplies by invading another cell, such as a bacterium or a body cell, and forcing it to make more copies of the virus. This kills the "host" cell. Diseases caused by viruses include the common cold, influenza, measles, mumps and rabies.

Some viruses are used in genetic engineering or modification—after they have been made harmless. Particularly useful are the viruses that attack bacteria, known as bacteriophages. They carry new genes into bacteria, and the bacteria then use the genes to make useful substances, such as medical drugs.

HIV (red), the cause of AIDS, breaks out from a human white blood cell.

 **FACT FILE**

THE AIDS VIRUS

AIDS, acquired immune deficiency syndrome, attacks the body's immune system. The immune system normally protects the body against diseases, especially those caused by invading bacteria and viruses. AIDS itself is caused by a virus, HIV, human immunodeficiency virus. Like certain other viruses the genes of HIV are in the form of RNA rather than DNA. Viruses that carry their genes as RNA rather than DNA are known as retroviruses.

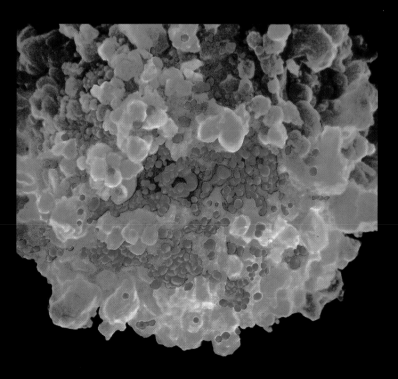

GENETIC ENGINEERING

↑ **The unique features of dog breeds like the basset hound have taken many generations to produce.**

Obtaining the Gene

First the desired gene's DNA is identified and "cut out" using chemicals called restriction enzymes. The same enzymes are used to cut a vector, which is a tiny bit of DNA (or RNA) that will eventually carry the gene to its new home. Types of vectors include plasmids and phages, which are viruses that attack bacteria.

Each type of restriction enzyme snips the DNA double helix at a certain place, between two specific base subunits. Since the desired gene DNA and the vector DNA were both cut with the same enzymes, their ends match. So the desired gene DNA can be joined or spliced into the "cut" in the vector DNA.

DNA Recombined

The vectors are put in laboratory containers with the cells, which will become the new homes for the gene. The vectors invade the cells and transfer the desired DNA, splicing it into the cell's own DNA so that it becomes part of the genome. The results are genes from different sources spliced together in new combinations, known as recombinant DNA. Living things with genes from other, different kinds of living things are called transgenic organisms.

Animals and plants have been genetically altered for thousands of years. People selected and bred them from wild ancestors and gradually made them tame or domestic and useful for food, transportation, hunting, and other purposes. The traditional method of selective breeding took many years, as different individuals were chosen or selected for each generation.

Genetic modification can now be carried out in just one generation by genetic engineering. This involves transferring genes from one type of living thing to another different type, using artificial methods in the laboratory.

FACT FILE

DNA FINGERPRINTING
DNA fingerprinting was developed in Britain in 1984. It relies on short, repeated sections of DNA found at many sites along human chromosomes. The number of repeats at a particular site varies between individual people. It can be measured from a sample of body tissue, even a tiny piece of hair or speck of blood. Samples from a crime scene can also be analyzed in the same way. Like real fingerprints, DNA fingerprints are almost unique to each individual. A match between the sample at the scene and the sample from a suspect means the suspect was almost certainly present at the scene.

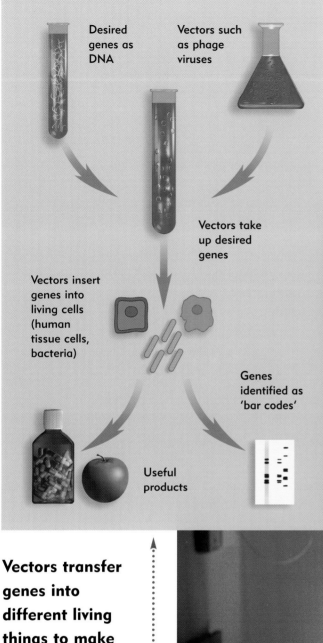

Desired genes as DNA

Vectors such as phage viruses

Vectors take up desired genes

Vectors insert genes into living cells (human tissue cells, bacteria)

Genes identified as 'bar codes'

Useful products

The Rise of Genetic Engineering

Genetic engineering has become big business. It can make extra copies of genes for further study, as in the Human Genome Project. However, a simpler technique called PCR, polymerase chain reaction, is now usually used for this. Genetic engineering is also used in biotechnology, medicine and cloning.

Vectors transfer genes into different living things to make useful products such as stay-fresh tomatoes and new medicines.

Lengths of DNA representing genes show up in chemical tests as "bar codes."

BIOTECHNOLOGY AND GMOs

Biotechnology involves growing cells for industrial purposes. In this sense, brewing (making beer) was one of the first biotechnologies, developed thousands of years ago. The alcohol in beer is produced by the activity of single-celled living things called yeasts.

Today most biotechnology involves bacteria or other cells that are modified by inserting other genes into them. The bacteria are grown or cultured in flasks or large vats. They produce protein substances according to the genes put into them. Some of these proteins are medically useful. The human gene for making the hormone insulin has been put into bacteria, which then make insulin. This is used to treat the medical condition of diabetes, which involves lack of insulin. Bacteria can also be modified with new genes for non-medical uses. These include the production of enzymes for washing powders, or bacteria with the ability to dissolve oil and so clear up polluting oil spills.

DNA samples are purified in a laboratory to identify their genes.

FACT FILE

JUMPING GENES
Genes seem to be strung along a chromosome like beads on a necklace. But studies carried out by US scientist Barbara McClintock in the 1940s showed that this was not quite accurate. Working with maize and other plants she discovered that some genes, especially those which control other genes, can "jump" from one site on their chromosome to another site, or even to another chromosome. In 1983, McClintock received a Nobel Prize for her discovery of these "transposable elements."

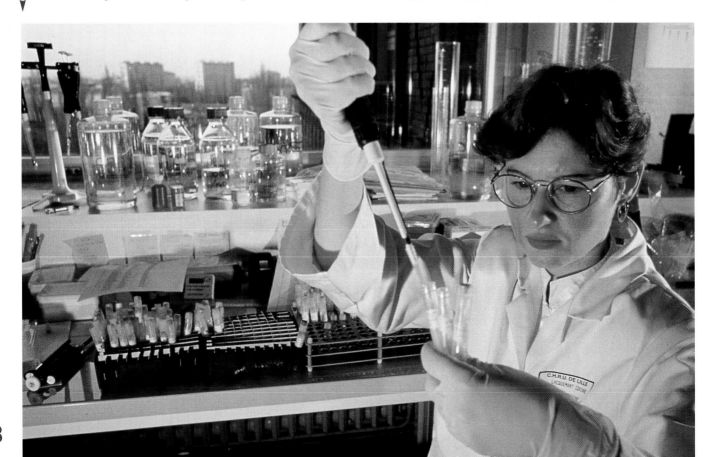

GM Life-forms

Genetic engineering enables genes to transfer directly between very different life-forms. Some GMOs (genetically modified organisms) could have tremendous benefits. GM crops and farm animals could be made that are more resistant than existing types to disease, drought, flood, or attack by pests. In the future this may help to ease famine and disease and increase food production around the world.

GM dangers

But there are many concerns about GM crops and other GMOs. The "foreign" genetic material in a GM crop might spread into other, similar plants in the wild and alter them in unknown ways. It could get into plant viruses and make them more powerful. Once in the environment, the genes might have all kinds of terrible and unforeseen effects—and they cannot be "put back" into their laboratory containers. Campaigns against GMOs urge great caution or even a halt to GM research until the risks are fully understood.

Genetic modification arouses strong opinions and emotions. Pieces of DNA put into farm crops or animals might "escape" into the environment with unknown consequences.

If genes for resistance to weedkillers were put into crops, weedkiller sprays might become more effective.

Where did human beings come from? Genetics can help to answer this age-old question. If DNA samples of humans are compared with other animals, the results show that the DNA of chimpanzees is more than 98 percent similar to ours. (That is, less than 1/50th of their DNA is different from ours.) We can also work backwards, knowing the average rate at which DNA changes or mutates in nature. This suggests humans and chimps both evolved from a common ancestor that lived about 5 million years ago. Fossils of human-like creatures dating back some 4 million years have been found in Africa, supporting this view.

Start in Africa

All people around the world belong to the same species, the modern human being, *Homo sapiens*. Studies of DNA in people from different regions suggest that the first modern humans appeared in Africa less than 200,000 years ago. Some time later— perhaps 100,000 years ago—groups of people started to spread from Africa around the world.

Can you roll your tongue into a tube shape? This ability depends on genes.

A chimp-like ancestor may have evolved into modern Homo Sapiens.

TEST FILE

ROLL YOUR TONGUE?
Try rolling up the sides of your tongue to form a U-shape or tube. Some people can, others cannot. It depends on your genes. However, lack of this curious ability is not likely to affect daily life!

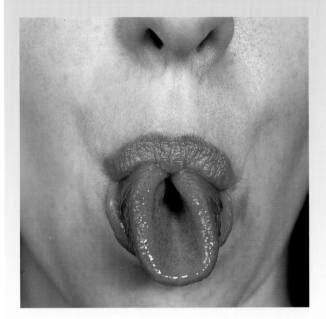

A family of chimpanzees

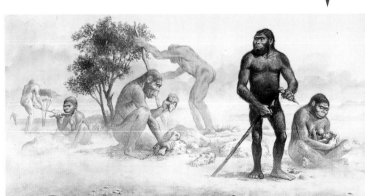

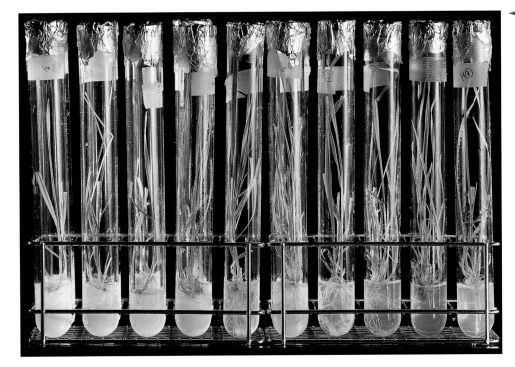

These cereal plants are all clones of one "parent" plant.

Making clones of animals, especially mammals and even humans, is more difficult. It is also very controversial. Each body cell contains a complete set of genes, but most of the genes are switched off. The task is to switch them on again so the single cell can activate all of its genes and, like a fertilized egg cell, develop into a new individual. This was achieved in 1996 to produce Dolly the Sheep. Cloning of human cells is the subject of great debate.

Genetics: Behaving and Thinking

The way people grow up is partly determined by the genes they inherit. But we are also affected by our environment—the climate where we live, the food we eat, our culture and customs, the schools we attend, and many other features of our surroundings. It is very difficult to untangle the effects of genes from those of the environment.

For example, research shows that several genes may control how memories are stored in the brain. This could mean that some people have a better natural ability to remember, compared to other people. But how we use our memory, and activities such as memory training or "learning to remember," can greatly change this inborn ability.

Cloning

Clones are living things that have exactly the same genes. In this scientific sense, "identical" twins have the same genes and so are clones of each other. Gardeners have made genetic copies or clones for thousands of years. They take cuttings of twigs or stems and plant them. The cuttings grow into new plants with the same genes as the single "parent" plant.

Dolly grew from a single cell of her "mother."

Clones of plants such as cereals are grown on a regular basis. Cloning animals, such as Dolly the Sheep, is a cause of much greater debate.

GENETICS AND MEDICINE

enetic conditions are those caused by abnormal genes. Some crop up unexpectedly, often caused by faulty cell division. Down's syndrome is due to an extra copy of a whole chromosome, number 21.

Many other genetic conditions are hereditary. They "run in families." Sometimes the way they occur is clear-cut because only one or a few genes are involved. For instance, if one parent has the condition or carries the gene for it, any children will have a predictable risk of inheriting it too.

In other cases many genes are involved. These conditions vary enormously, from slight asthma to Alzheimer's disease. The risks of children developing the condition, and how severe it may become, are much more difficult to predict.

However, in one sense, all diseases are partly genetic. Even if there is a non-genetic cause, such as infection by germs, an individual's genes affect how the body reacts to the disease.

Testing for Genetic Conditions

People can be tested or screened for certain genetic conditions. The DNA from tiny samples of body tissues is analyzed in the laboratory for faulty or missing genes. An expert called a genetic counselor can then advise the person on any risks involved. However, there is great concern about who should have access to such information.

Tests can indicate the presence of cystic fibrosis, a hereditary disease.

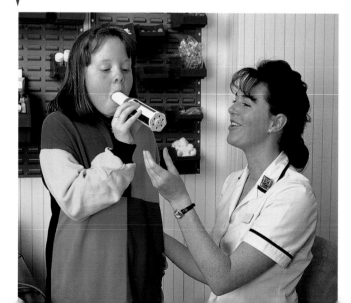

FACT FILE

SICKLE-CELL ANEMIA

In this disease, red cells in the blood become sickle-shaped—bent or curved. It causes pain, lack of breath, and other problems. The condition is due to one changed gene. If a person receives two copies of this gene, one from each parent, he or she develops the condition. A person with one changed and one normal version is hardly affected but can pass on the changed gene to his or her children.

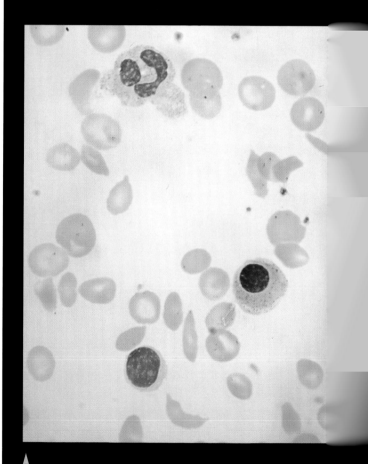

Some of the red blood cells in this sample (magnified about 1,100 times) are curved, indicating the blood condition sickle-cell anemia.

A technician uses a gene gun to inject DNA into the cells of plant leaves. The DNA may enter the cell nuclei and give the plant genes for new features.

FUTURE FILE

SPARE-PART CLONES

Cloning—creating genetically identical copies—may help people with certain diseases. Cells are taken and the genes altered to remove the problem. The cells are then grown in the laboratory to form new tissue. For example, heart cells could be grown into new heart muscle tissue. The tissue is then put back into the person to cure the problem. Since the cells came originally from that person, there should be no problem of rejection as may happen in a normal transplant. The body tries to reject or get rid of the transplant since it is genetically different.

Gene Therapy

Gene therapy (gene transplant) involves genetic engineering to replace faulty genes with normal, healthy versions. However, most cells in the body have a limited existence, usually hours or days. They are made by cell division, carry out their tasks, die, and are replaced as part of normal body maintenance. If their genes are changed, the effects would only last a limited time.

New Genes for Old

One aim of gene therapy is to replace the faulty genes in so-called stem cells. These are "ancestor cells"—fairly unspecialized cells that keep dividing in order to make slightly more specialized cells, which then divide to produce fully specialized cells. Stem cells, in effect, survive for years or even the individual's lifetime. Changing their genes would mean all their "descendants" receive healthy genes, so the effects could last much longer.

Such new techniques still have technical problems. But they have been used successfully on babies suffering from a rare disorder of the immune system.

A doctor and a genetic counselor advise potential parents on the risks of passing inherited conditions to their children.

GLOSSARY

Allele A form or version of a gene. For example, the gene for eye colour may have red, brown, grey, green and other alleles.

Cell A microscopic living unit—a "building block" of life. Some living things are single cells. Others are made up of thousands or millions of cells.

Chromatid One of a pair of duplicate chromosomes formed by DNA replication (copying), visible when a cell is about to split or divide.

Chromosome A thread-like structure inside a living cell, made mainly of DNA and proteins. It carries genetic information for that living thing and is most easily visible as the cell is about to divide.

Diploid Having two copies or versions of each chromosome, like normal body cells. (See also *haploid*.)

DNA Deoxyribonucleic acid, the chemical substance that carries genetic information in the vast majority of living things.

Dominant When one version or allele of a gene is more powerful than another (which is recessive) and makes its full effect felt.

Gene In general terms, a unit of inheritance for passing on one feature, characteristic or trait. For example, there is a gene for eye color. However in modern genetics a gene is described as a sequence of DNA (or RNA) that carries the coded information, in chemical form, to make one protein.

Genome The full set of genes for a living thing.

Haploid Having one copy or version of each chromosome, like eggs and sperm cells. (See also *diploid*.)

Heredity Passing on features from parents to offspring, usually by means of genes.

Karyotype All of the chromosomes from a living thing—the full set. In humans this is 46 (23 pairs).

Recessive When one version or allele of a gene is less powerful than another (which is dominant) and cannot make its effect felt.

RNA Ribonucleic acid, a chemical substance that can carry genetic information and also transfer genetic information from DNA to build proteins.

Species A group of similar living things, whose members can breed with each other to produce more of their kind, but which cannot breed with other species.

FURTHER INFORMATION

FURTHER READING

Balkwill, Fran. *Cell Wars*. Lerner Publications, 1994.

Darling, David. *Genetic Engineering: Redrawing the Blueprint of Life*. (Beyond Two Thousand series). Silver Burdett Press, 1995.

Klare, Roger. *Gregor Medel: Father of Genetics*. (Great Minds of Science series). Enslow Publications, 1997.

Marsh, Carole. *Is One Enough? Are Two Too Many?* (Cloning for Kids! series). Gallopade International, 1998.

Snedden, Robert. *The History of Genetics*. (Science Discovery series). Raintree Steck-Vaughn, 1995.

Wilcox, Frank H. *DNA: The Thread of Life*. (Discovery! series). Lerner Publications.

PLACES TO VISIT AND WEB SITES

Genetic Science Learning Center
Eccles Institute of Human Genetics
University of Utah
Salt Lake City, UT 84112
http://gslc.genetics.utah.edu/

Science Museum of Virginia
2500 West Broad Street
Richmond, VA 23220
http://www.smv.org/prog/treelife.htm

Human Chromosome Launchpad
www.ornl.gov/hgmis/launchpad/
Information about each human chromosome with links to gene maps, genetic disorders, identified genes and DNA research.

Human Genome Project Information
www.ornl.gov/TechResources/Human_Genome/home
The main homepage for the Human Genome Project.

The Genome Database
www.gdb.org/
Latest technical news on the human genome.

Rare Genetic Diseases In Children Homepage
http://mcrcr2.med.nyu.edu/murphp01/

INDEX

ENTIRE
CHROMOSOME

Centromere

Super-supercoils
packaged with
proteins called
histones

Supercoils of
DNA

Backbone of DNA consists
of repeated deoxyribose
sugar and phosphate
subunits

Coils of DNA

Chromosomes and DNA

A chromosome may consist of a hugely long but
incredibly thin piece, or molecule, of DNA
(deoxyribonucleic acid). DNA has a double-helix
structure like a ladder twisted into a corkscrew
shape. Its genetic information is carried as a
chemical code of four subunits called bases—A, T,
G and C. The double-helix is twisted into coils, and
these coils are looped into supercoils, and so on.
Such coiling fits the piece of DNA, which may be
several centimeters long, into a chromosome less
than 1/100th of a millimeter in length.

Bases form cross-links
and form genetic code

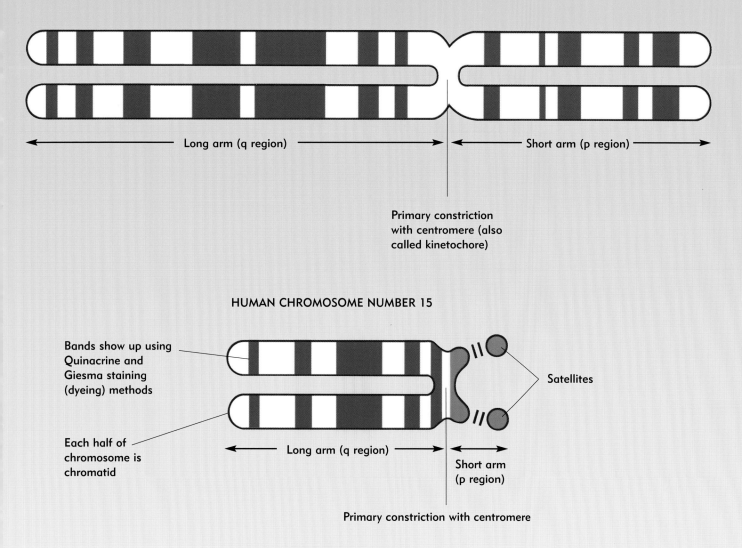

Long arm (q region) ◄———————————————————► Short arm (p region)

Primary constriction
with centromere (also
called kinetochore)

HUMAN CHROMOSOME NUMBER 15

Bands show up using
Quinacrine and
Giesma staining
(dyeing) methods

Each half of
chromosome is
chromatid

Satellites

◄——— Long arm (q region) ———► Short arm
(p region)

Primary constriction with centromere

Studying Chromosomes

It is not possible to see genes or individual sequences of DNA on
a chromosome, because the DNA is far too thin and tightly coiled.
But under the light microscope, chromosomes can be dyed or
stained to reveal patterns of stripes or bands along their length.
The banding varies among chromosomes in the full set and helps
to identify each one. Unusual banding patterns show where pieces
of chromosome have broken off, gone missing or even been
swapped with bits of other chromosomes—a process called
crossing-over. Human chromosomes are numbered from 1, the
longest, to 22, the shortest, plus the sex chromosomes X or Y.